THE INTERNET SAFETY HANDBOOK

INVESTIGATE!

OUTSMART COMPUTER VIRUSES

BY CHARLEY LIGHT

Enslow PUBLISHING

Please visit our website, www.enslow.com. For a free color catalog of all our high-quality books, call toll free 1-800-398-2504 or fax 1-877-980-4454.

Library of Congress Cataloging-in-Publication Data
Names: Light, Charley, author.
Title: Outsmart computer viruses / Charley Light.
Description: Buffalo, New York : Enslow Publishing, [2024] | Series: The Internet safety handbook | Includes bibliographical references and index.
Identifiers: LCCN 2023009594 | ISBN 9781978535015 (library binding) | ISBN 9781978535008 (paperback) | ISBN 9781978535022 (ebook)
Subjects: LCSH: Computer viruses–Juvenile literature. | Computer networks–Juvenile literature. | Computer security–Juvenile literature.
Classification: LCC QA76.76.C68 L54 2024 | DDC 005.8–dc23/eng/20230310
LC record available at https://lccn.loc.gov/2023009594

Published in 2024 by
Enslow Publishing
2544 Clinton Street
Buffalo, NY 14224

Portions of this work were originally authored by John M. Shea and published as *Combating Computer Viruses*. All new material in this edition authored by Charley Light.

Designer: Jen Schoembs
Editor: Charley Light

Photo credits: Cover, p. 1 (girl) Prostock-studio/Shutterstock.com; cover, p. 1 (background), series art (background) kerly chonglor/Shutterstock.com; p. 5 fizkes/Shutterstock.com; p. 7 pinkeyes/Shutterstock.com; p. 9 niroworld/Shutterstock.com; p. 11 Feng Yu/Shutterstock.com; p. 13 Zai Di/Shutterstock.com; p. 15 Rawpixel.com/Shutterstock.com; p. 17 asharkyu/Shutterstock.com; p. 19 Evan Lorne/Shutterstock.com; p. 21 Andrey_Popov/Shutterstock.com; p. 23 IhorL/Shutterstock.com; p. 25 Nopparat Khokthong/Shutterstock.com; p. 27 Max Morphine/Shutterstock.com; p. 29 NicoElNino/Shutterstock.com.

Printed in the United States of America

CPSIA compliance information: Batch #CSENS24: For further information contact Enslow Publishing at 1-800-398-2504.

CONTENTS

Words in the glossary appear in **bold** type the first time they are used in the text.

GET READY TO SAFELY SURF

The internet can be a magical place to learn and connect with others. It also has certain dangers to be aware of. Fortunately, knowledge is a great tool for self-defense, and you've already armed yourself with this book!

This book will introduce you to the basics of malicious software that lurks online. Malicious means hurting someone or something on purpose. Software is the programming on digital devices that tells those devices what to do. You'll also learn some ways to protect your digital devices from malicious software.

Make a plan with your guardian about how to stay safe online.

EXPLORE MORE

Have you ever had ads pop up on your screen? Ads like this are called adware. Adware is annoying and sometimes unsafe, but not necessarily malicious. Software like this is called "potentially unwanted programming," or PUPs for short.

VICIOUS VIRUSES

In **biology,** a virus is a tiny particle that makes copies of itself. This kind of virus needs a **host** to live inside of. A computer virus works the same way. It's a computer program that makes copies of itself. It needs a computer or other digital device as a host.

Computer viruses are one kind of malware, which is short for "malicious software." Malware is unwanted programming placed on a person's computer without their permission or knowledge. Malware often enters a digital device by hiding in another program or file, such as a photo.

EXPLORE MORE

Viruses can slow or shut down computers and the **networks** that connect computers. The Slammer virus caused many computer networks to crash in 2003, including the 911 services of Seattle, Washington. Viruses cost both businesses and families billions of dollars each year.

Computer viruses were named after viruses like the flu, shown here, because they act so similarly.

A VIRAL LIFE

A computer virus is made of different parts. The **infection** mechanism is how a virus spreads itself. It includes a search routine, which looks for host programs for the virus to infect. It also has a copy routine, which is how the virus copies itself once inside a host.

The trigger is the part that activates, or wakes up, the virus. The payload is the part holding the harmful work the virus was built to do. The anti-detection routine is the part that hides the virus from being detected, or found.

LIFE CYCLE OF A TRADITIONAL COMPUTER VIRUS

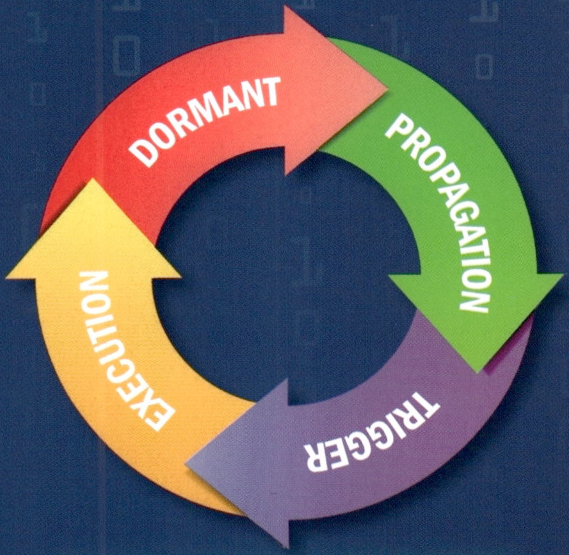

Dormant

The virus doesn't do anything, so the user doesn't know their computer has been infected yet. Some viruses skip this stage.

Propagation

The virus propagates, or copies, itself. The virus puts each copy into another place on the infected device. Each copy starts its own propagation phase.

Trigger

The trigger is an event such as a certain amount of time passing. It activates, or wakes up, the virus.

Execution

The payload is executed, or released. The harmful work begins.

Some viruses go through these stages one time. Others cycle through the stages.

EXPLORE MORE

Some viruses morph, or change, their copies. Biological viruses can do this too! It's a survival mechanism, or a tool that helps viruses keep going. Morphed copies of the virus are all different. This makes the copies harder to detect—or stop.

YOU'VE GOT MAIL!

Email malware attacks happen through—you guessed it—email. Email malware often hides in attachments. These **attachments** are usually **executable files**. Common executable files have names ending in .exe, .com, or .pdf. Malware can also hide in document files and photos.

Emails can contain links to harmful websites that download malware to computers secretly. In 2007, many people received emails with the subject "230 Dead as Storm Batters Europe." When they clicked on a link in the email for more information, they were taken to a website that downloaded the Storm malware to their computer. Millions of computers were infected!

EXPLORE MORE

Malware can spread through instant messaging apps such as Discord as well. These mostly spread through infected links and infected files shared in the app. Some spread through infected Discord installation files—the file you download to access Discord in the first place.

Email services sort mail that looks dangerous into a folder called "spam" or "junk."

THE WORMS CRAWL IN

Malware worms don't hide inside files and they don't need a trigger. They just attack. When a worm infects a network, it looks for other computers on that network to infect. The worm then copies itself onto the computer and continues to search for more uninfected computers.

Worms often spread by email attachments in order to infect many networks at once. However, you don't need to download or open anything to become infected by a worm. Just having a computer connected to a network, including the internet, increases the risk of a worm infection.

EXPLORE MORE

The MyDoom worm was so big that it took down an entire search engine, Google, and slowed several more. When the attack started in 2004, as many as one in 10 emails in the world was connected to this worm.

In 2000, the ILOVEYOU worm sent emails with the subject "ILOVEYOU" to all of a person's email contacts to spread itself.

TROJAN HORSES

Trojan horses are malware named from Greek mythology. During a war with Troy, the Greeks built a giant wooden horse and left it outside the city gates. The Trojans thought it was a gift and took the Trojan horse into their city—along with the Greek soldiers hiding inside it!

A Trojan horse malware works the same way. It's a dangerous program that masks itself as something innocent, such as a free game or movie. People are tricked into downloading a Trojan horse. While this malware doesn't copy itself, it can still damage, or harm, devices.

EXPLORE MORE

The Zeus Trojan malware was built to steal from banks. Once it has infected a device, it watches the websites users go to—including online banks. It also records the keystrokes users enter—including their bank passwords! The malware's owner can then steal money.

Malware that spies on people, like the Zeus Trojan, is called spyware.

WALL OF FIRE

If you're going to connect your computer to the internet, you need a trusted firewall. A firewall is a security device that acts as a barrier between a computer and the network. It watches information moving in and out of the computer.

A good firewall will allow safe information in, such as honest websites and emails, and keep malware out. Firewalls also stop people from outside your private network from logging onto your computer. But a firewall alone will not protect you from viruses. You also need antivirus software.

Firewalls can be hardware (physical devices) or software (programs).

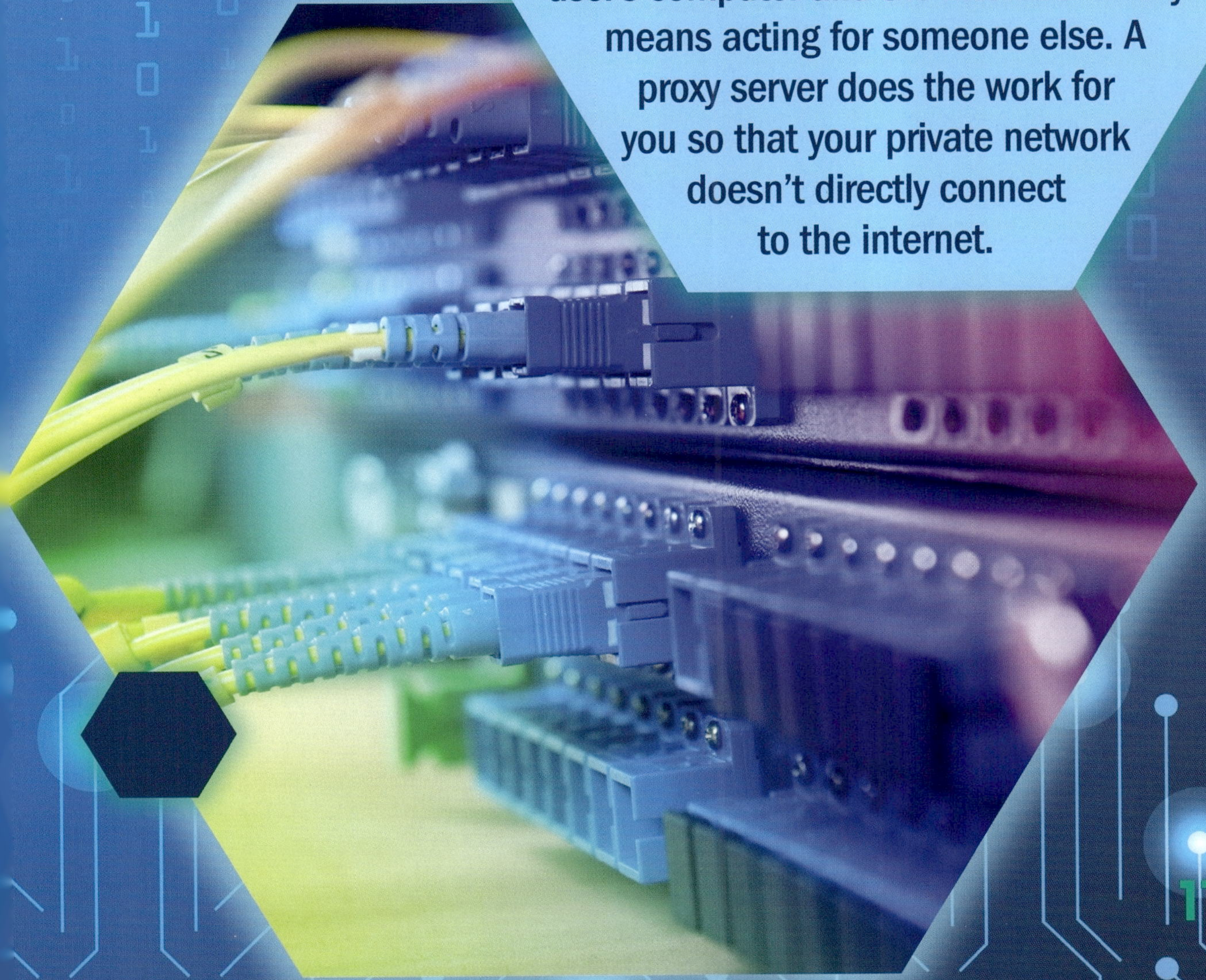

EXPLORE MORE

A proxy server can be part of a firewall. It's another barrier between the user's computer and the internet. "Proxy" means acting for someone else. A proxy server does the work for you so that your private network doesn't directly connect to the internet.

ANTIVIRUS PROGRAMS

Antivirus programs protect your devices from viruses and other malware. There are many antivirus programs to choose from. Some of the best require users to pay for it. However, other good antivirus programs are free.

Antivirus programs search inside each file on a computer for known viruses and malware. They also watch programs for suspicious activity, such as making copies or trying to access the internet secretly. If an antivirus program finds this activity, it will stop the suspicious program and alert the user.

Web browsers like Firefox have extensions, or add-ons, that protect you from malicious websites too.

EXPLORE MORE

For best results, an antivirus program should run all the time. Then, it can catch viruses before they have a chance to copy themselves onto a hard drive. Nowadays most antivirus programs update themselves automatically so they're ready for the latest threats.

GONE PHISHING

Phishing means pretending to be something innocent in order to trick users into downloading a virus or malware. Lots of malware is spread through phishing emails and websites.

The Dharma malware is hidden in phishing emails. The email pretends to be from Microsoft and tells users to download an antivirus program. When users click the link, they download real antivirus software and Dharma malware at the same time! While the user is busy installing the real antivirus software, the Dharma malware encrypts the user's information.

EXPLORE MORE

Encryption turns information, such as documents, into a secret code. Ransomware such as Dharma encrypts a person's information. They have to pay a ransom (usually money) to the ransomware's owner to get a key to decrypt their information again.

Phishing is the number one way ransomware spreads.

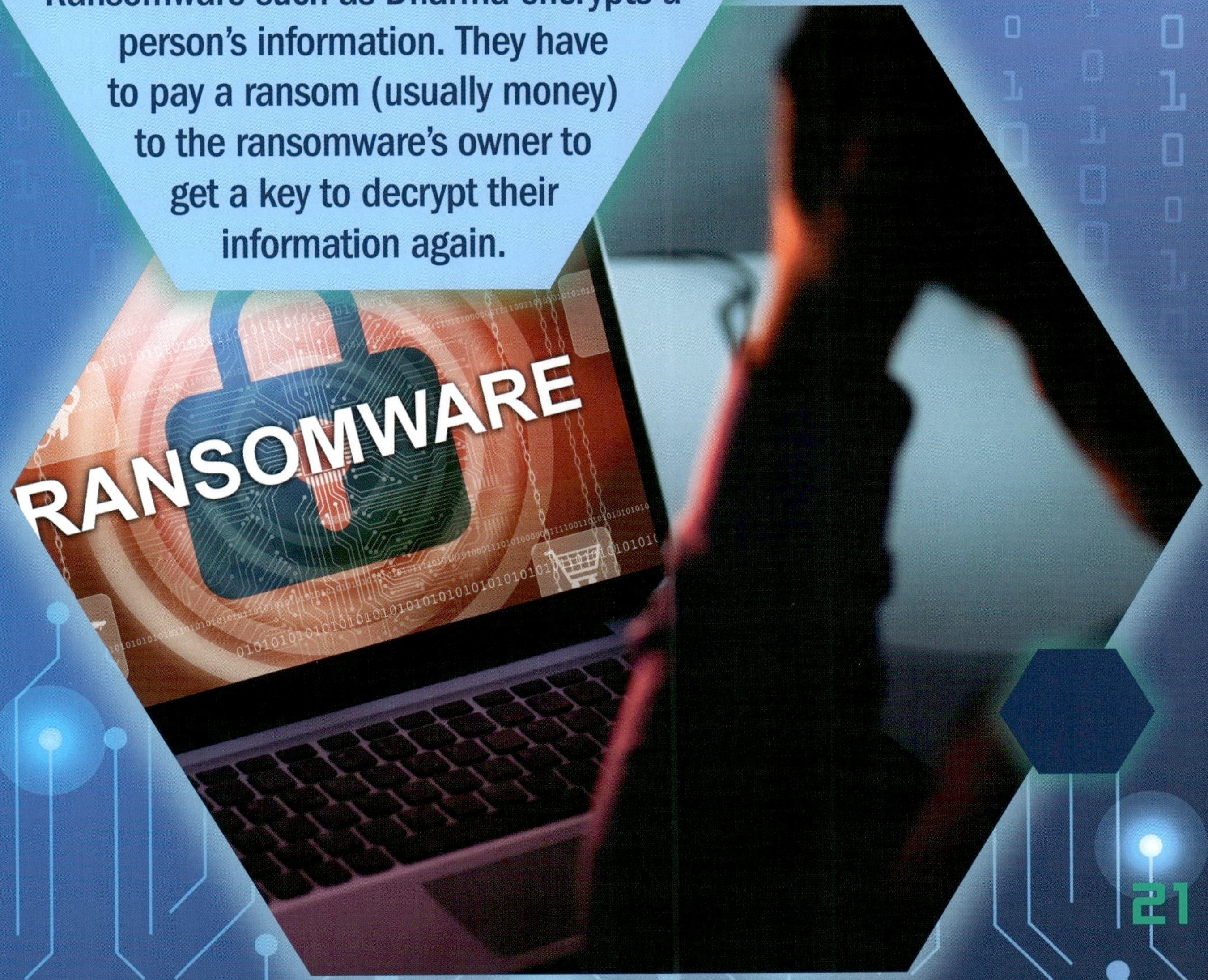

BACK IT UP

Any device that connects to other networks, including the internet, is at risk of malware. Even with the best antivirus programs, the strictest firewall, and the most careful user, there's always a chance a computer will be infected with a virus or other malware.

By making copies of important files, you can reduce the damage it can cause. Backing up, or making copies of files, must be done ahead of time, before a computer is infected. Backups keep your data, or information, safe from things such as **disasters** or theft too.

EXPLORE MORE

USB flash drives are a popular backup storage system. They're also called thumb drives or USB sticks. Unfortunately, they can also be used to spread malware from device to device as the stick is plugged into different computers.

Never put an unknown USB in your device! It could be carrying malware.

HEAD IN THE CLOUD

Cloud storage systems are run online. The data you upload to the cloud is actually stored on a physical computer owned by the cloud storage company. You can access your data through the internet. If something happens to your devices at home, anything you've stored in the cloud will still be safe.

Cloud storage isn't perfect, either, though. Some cloud storage is free, but sometimes it costs money. Sometimes the cloud storage company loses power and users can't access their data. It's also possible for clouds to be **hacked**.

Some cloud storage systems such as Dropbox (below) use encryption so that you can read your files, but no one else can.

EXPLORE MORE

To prevent your backups from being infected—or destroyed by things such as fires—it's best to use multiple backup systems at once. It can be smart to use a combination of physical and cloud backup storage.

FAMOUS CASES

Some viruses and malware are mostly harmless. They may make annoying pop-ups (messages or pictures that open themselves on your computer). Others can have a huge impact on the world.

The Slammer worm infected 75,000 computers in 2003. It got to most of them within only 10 minutes! It slowed the internet down, shut down 911 services, and crashed the Bank of America's **ATMs**. A newer threat is clop ransomware. It can shut down the built-in cybersecurity programs on Windows computers and 600 other programs that might stop it!

PROTECT YOURSELF CHECKLIST

- ✓ make strong passwords
- ✓ install latest updates on all software
- ✓ use an antivirus software program
- ✓ use a firewall
- ✓ don't open attachments or links in emails from unknown senders
- ✓ don't use USBs or storage devices you don't know

EXPLORE MORE

Some malware waits for a certain time to attack. The Jerusalem virus erased a person's computer files every Friday the 13th. It is one of the oldest computer viruses, first discovered in 1987!

SO LONG AND THANKS FOR ALL THE PHISH

If you found this book interesting, you might like a career in cybersecurity. Security software developers make antivirus programs and other software to keep users safe. Security engineers build cybersecurity systems that can handle the needs of specific clients. This includes installing software.

You could even be an ethical hacker! These are people who are paid to test security systems by seeing if they can hack into them. This helps point out the flaws in a cybersecurity system that need to be fixed.

EXPLORE MORE

Rootkits are a way to access a computer remotely, or from a distance. They are used by IT (information technology) professionals to help people fix problems with their computers. They can also be used maliciously to take over someone else's computer.

Today, you can learn ethical hacking at many colleges!

GLOSSARY

ATM: A machine that performs basic banking functions, such as cash withdrawals.

attachment: File attached to an email or instant message.

biology: The study of living things.

disaster: A sudden event that causes great harm.

executable file: File that makes a program run upon being clicked.

hack: To access by illegal means

host: A living thing that is home to a parasite, or another living thing that harms its host to survive. It can also mean the computer or program a computer virus infects.

infection: The spread of something malicious through a device.

network: System of connected things.

FOR MORE INFORMATION

BOOKS

Holmes, Kirsty. *Hack Attack!* St. Catharines, ON: Crabtree Publishing Company, 2020.

Oxlade, Chris. *Computer Science for Curious Kids.* London, UK: Arcturus, 2023.

WEBSITES

Computer Virus Facts for Kids
kids.kiddle.co/Computer_virus
You can learn more about the stages of a computer virus attack here!

Dinopass
www.dinopass.com
This site generates passwords to keep your accounts safe.

Funbrain
www.funbrain.com
You can play fun educational games without downloading anything here!

Publisher's note to educators and parents: Our editors have carefully reviewed these websites to ensure that they are suitable for students. Many websites change frequently, however, and we cannot guarantee that a site's future contents will continue to meet our high standards of quality and educational value. Be advised that students should be closely supervised whenever they access the internet.

INDEX

MACHINES MADE SIMPLE

Pulleys

Written by E.C. Andrews

KidHaven
PUBLISHING

Published in 2026 by
KidHaven Publishing, an Imprint of
Greenhaven Publishing, LLC
2544 Clinton St., Buffalo, NY 14224

Written by: E.C. Andrews
Edited by: Elise Carraway
Designed by: Amelia Harris

Cataloging-in-Publication Data
Names: Andrews, E. C. (Children's author)
Title: Pulleys / E.C. Andrews.
Description: Buffalo, NY : KidHaven Publishing, 2026. | Series: Machines made simple | Includes glossary and index.
Identifiers: ISBN 9781534550759 (pbk.) | ISBN 9781534550766 (library bound) | ISBN 9781534550773 (ebook)
Subjects: LCSH: Pulleys--Juvenile literature. | Simple machines--Juvenile literature. | Hoisting machinery--Juvenile literature.
Classification: LCC TJ1103.A62 2026 | DDC 621.8'2--dc23

Manufactured in the United States of America

CPSIA compliance information: Batch #CSKH26
For further information contact Greenhaven Publishing LLC at 1-844-317-7404.

Please visit our website, www.greenhavenpublishing.com.
For a free color catalog of all our high-quality books, call toll free 1-844-317-7404 or fax 1-844-317-7405.

Find us on

Image Credits
All images are courtesy of Shutterstock.com, unless otherwise stated. Cover – King Ropes Access, Zerbor, CKYN stock photo, ComicSans. Recurring images – Onica Alexandru Sergiu. 2–3 – P24_Design, Heavypong, Konstantin Ivshin. 4–5 – GreenThumbShots, Pixel-Shot, Maria Sbytova. 6–7 – Margo Harrison, NassornSnitwong. 8–9 – AlyoshinE, Fer Gregory, watcher fox. 10–11 – Zerbor, Marco Fo. 12–13 – Chaitanya 7, Roy Hinchliffe. 14–15 – BearFotos, MUHAMMEDKARSLI. 16–17 – Konstantin Ivshin, kaninw. 18–19 – Olesia Bilkei, Roman Chekhovskoi, Yuliya Evstratenko. 20–21 – Harmattan Toujours, Stephen Barnes, Anton Balazh. 22–23 – Nach-Noth, Victoria Strelets.

Contents

WORDS THAT LOOK LIKE THIS CAN BE FOUND IN THE GLOSSARY ON PAGE 24.

Marvelous Machinery

We use machines every day. Machines can be very complicated and have lots of moving parts. When you think about machines, you may think about computers or big machines that are used to build things.

INSIDE A COMPUTER

Some machines are much simpler. Simple machines can be useful on their own. They can also be used in more complicated machines. Did you know that doors, wheelbarrows, and blinds work using simple machines?

MOST MACHINES ARE MADE UP OF SEVERAL SIMPLE MACHINES.

What Is a Machine?

Machines are usually made to do tasks that would be difficult or even impossible for people to do on their own. Machines can be used to lift heavy things or move them from place to place.

THIS CRANE IS ABLE TO LIFT A WHOLE BUILDING!

Simple machines affect the strength or direction of a force. Simple machines mostly do tasks by using push and pull forces. Hammers, knives, and screws are all examples of simple machines.

THIS HAMMER IS USING A PUSH FORCE ON A NAIL.

What Is a Pulley?

A pulley is a simple machine. Pulleys are often used to lift objects that are too big or heavy for a person to lift on their own, or too high for a person to reach.

Pulleys can also be used to open and close things. Pulleys used for opening and closing are often found on big, heavy structures. They are also used on things that are high up.

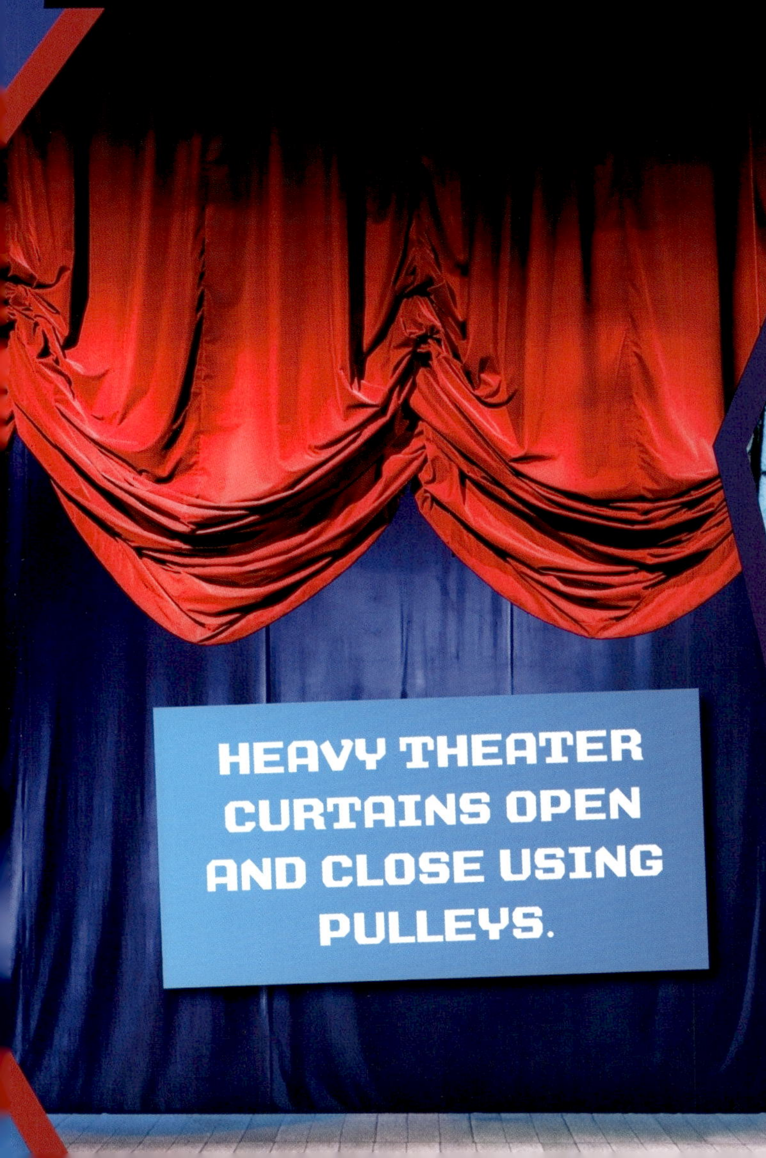

HEAVY THEATER CURTAINS OPEN AND CLOSE USING PULLEYS.

Parts of a Pulley

Pulleys have several parts that make them work.

Rope

Pulleys use ropes, chains, or cords. They help connect the objects that need lifting to the pulley.

Wheel

One of the most important parts of a pulley is the wheel. The wheel of a pulley has a groove in it. The rope sits in the groove of the wheel.

Axle

The axle is a cylinder-shaped rod that is placed through a hole in the middle of the wheel. This holds the wheel in place. It also allows the wheel to turn.

Frame

The frame is built around the pulley. The frame holds the pulley system in place. This gives the pulley system something strong and steady enough to pull an object toward.

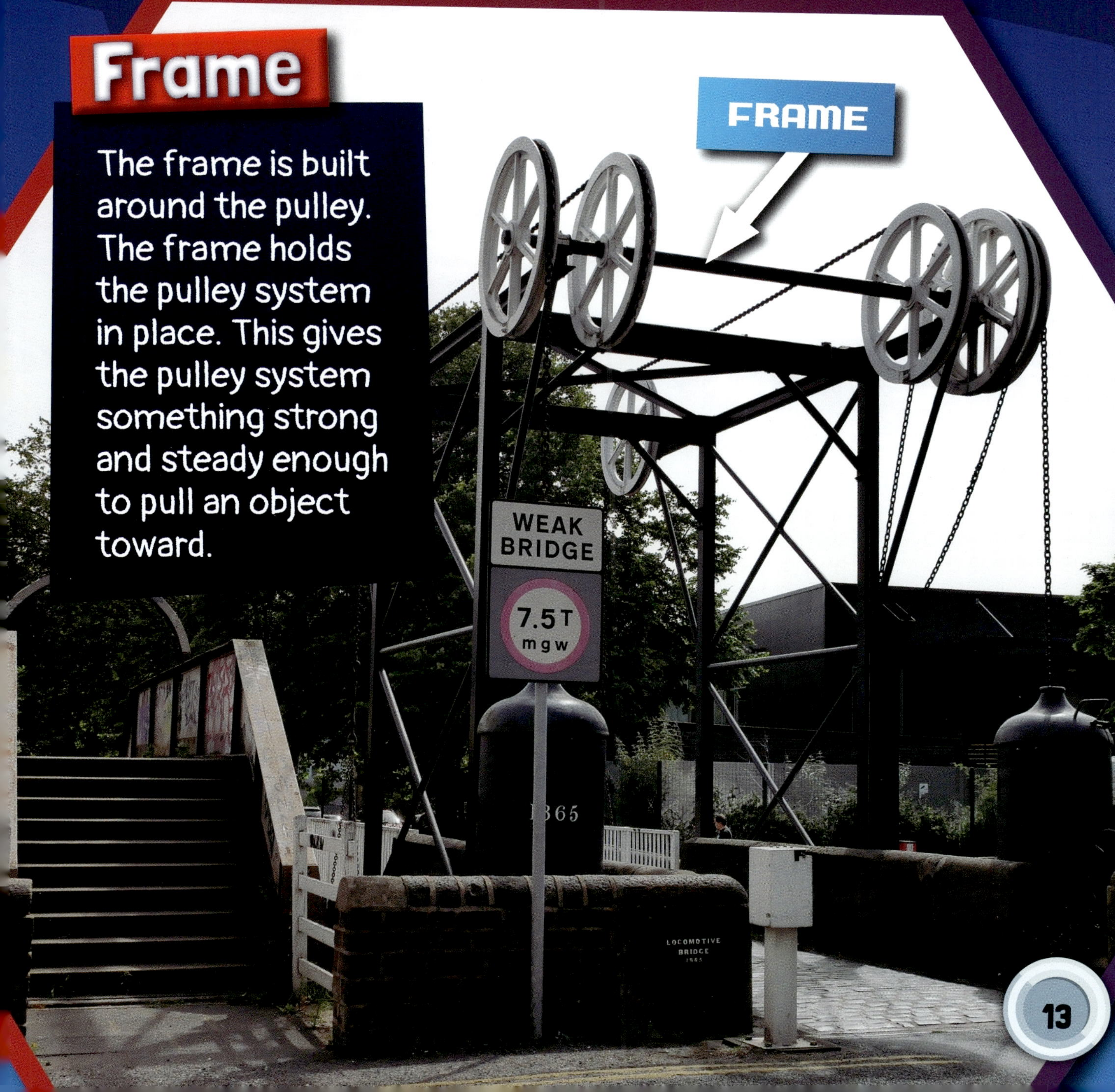

Working with Pulleys

Pulleys work by changing the direction of the force needed to lift an object.

Sometimes, pulleys can make it easier to lift heavy things.

Fixed pulleys do not move. They help change the direction of a pull.

Movable pulleys do move. They make things easier to lift.

Compound pulleys have both fixed and movable pulleys.

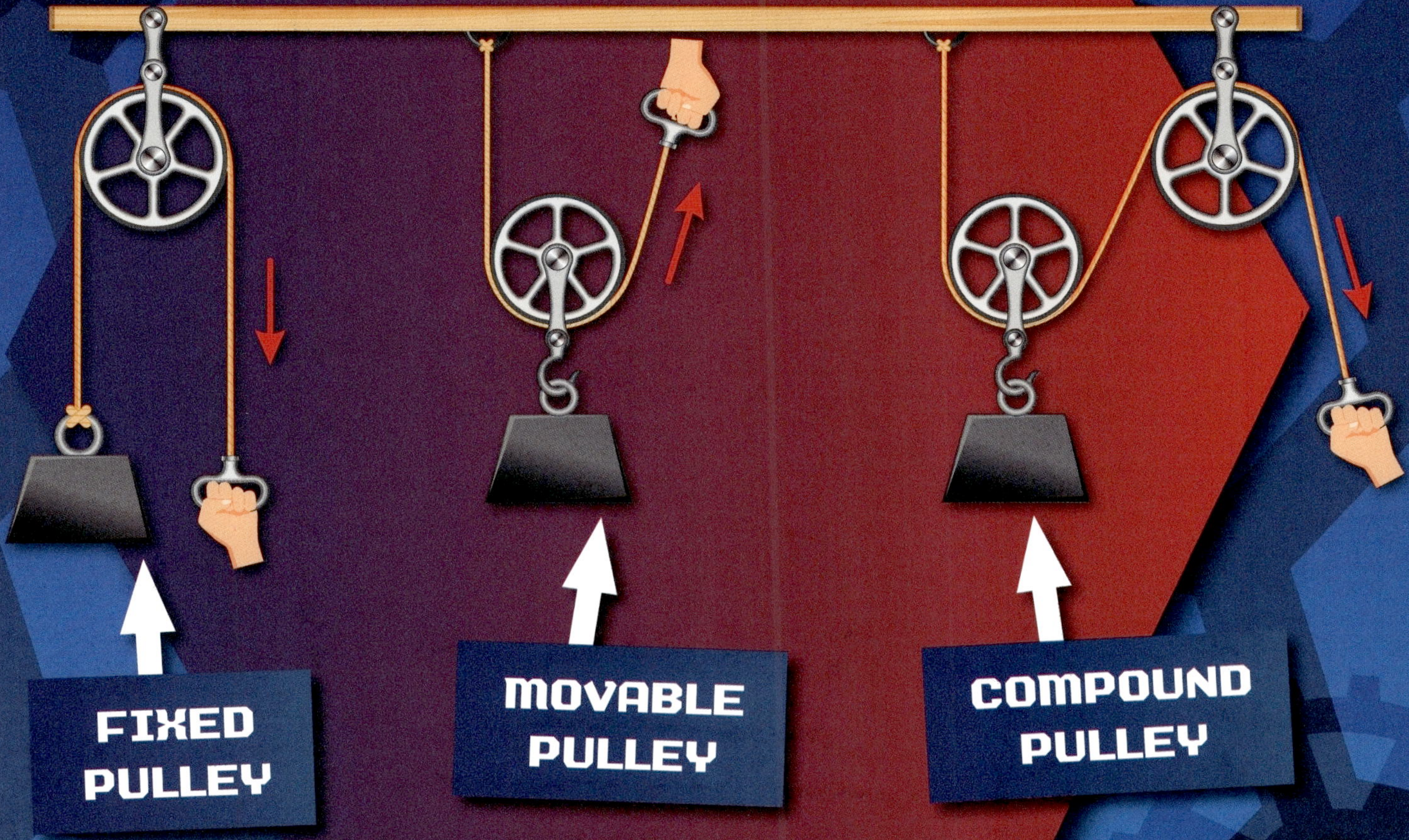

Practical Pulleys

Cranes

Cranes use movable pulleys. Cranes can lift very heavy objects up and down. Cranes are usually used to lift and move heavy materials on building sites.

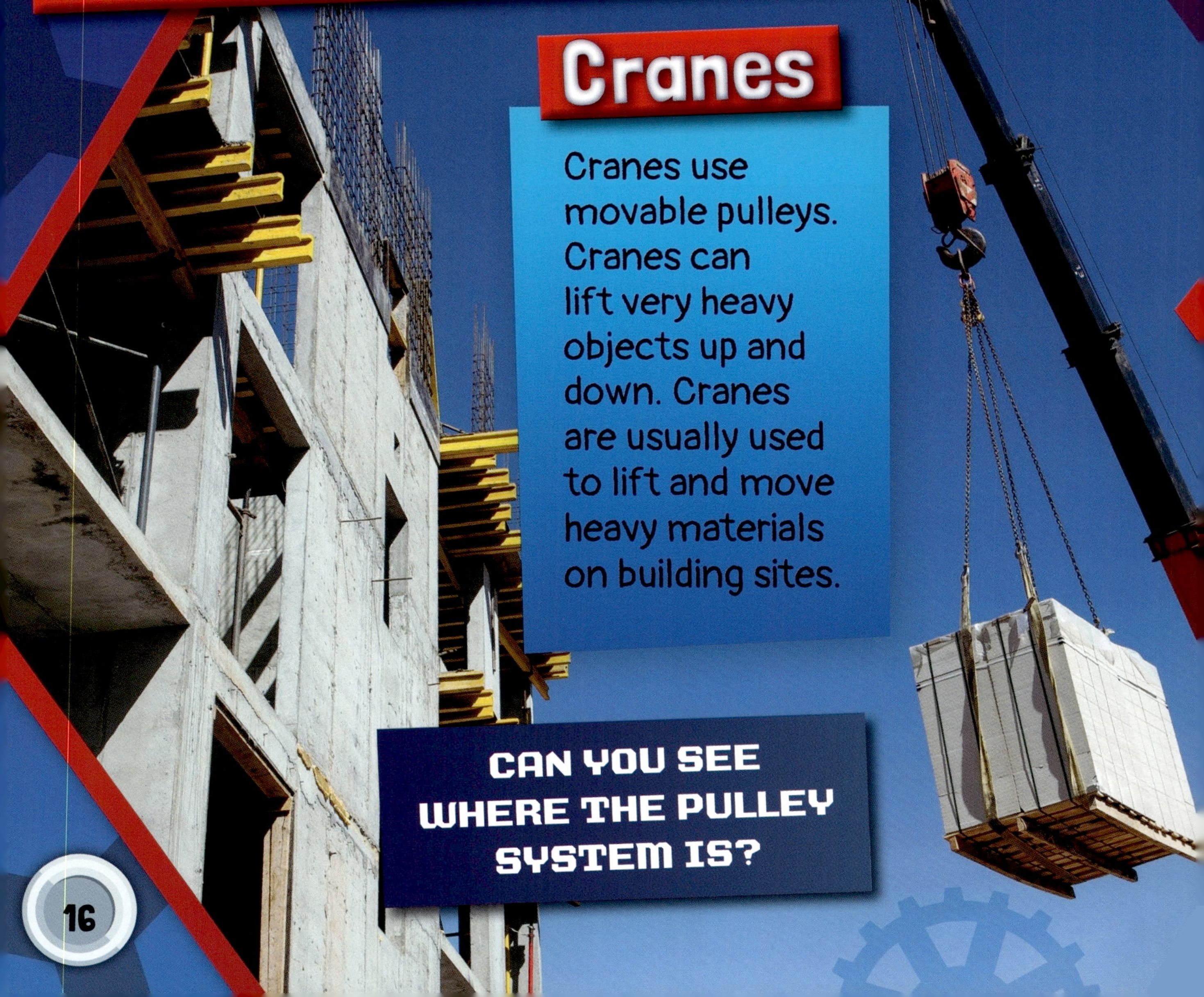

CAN YOU SEE WHERE THE PULLEY SYSTEM IS?

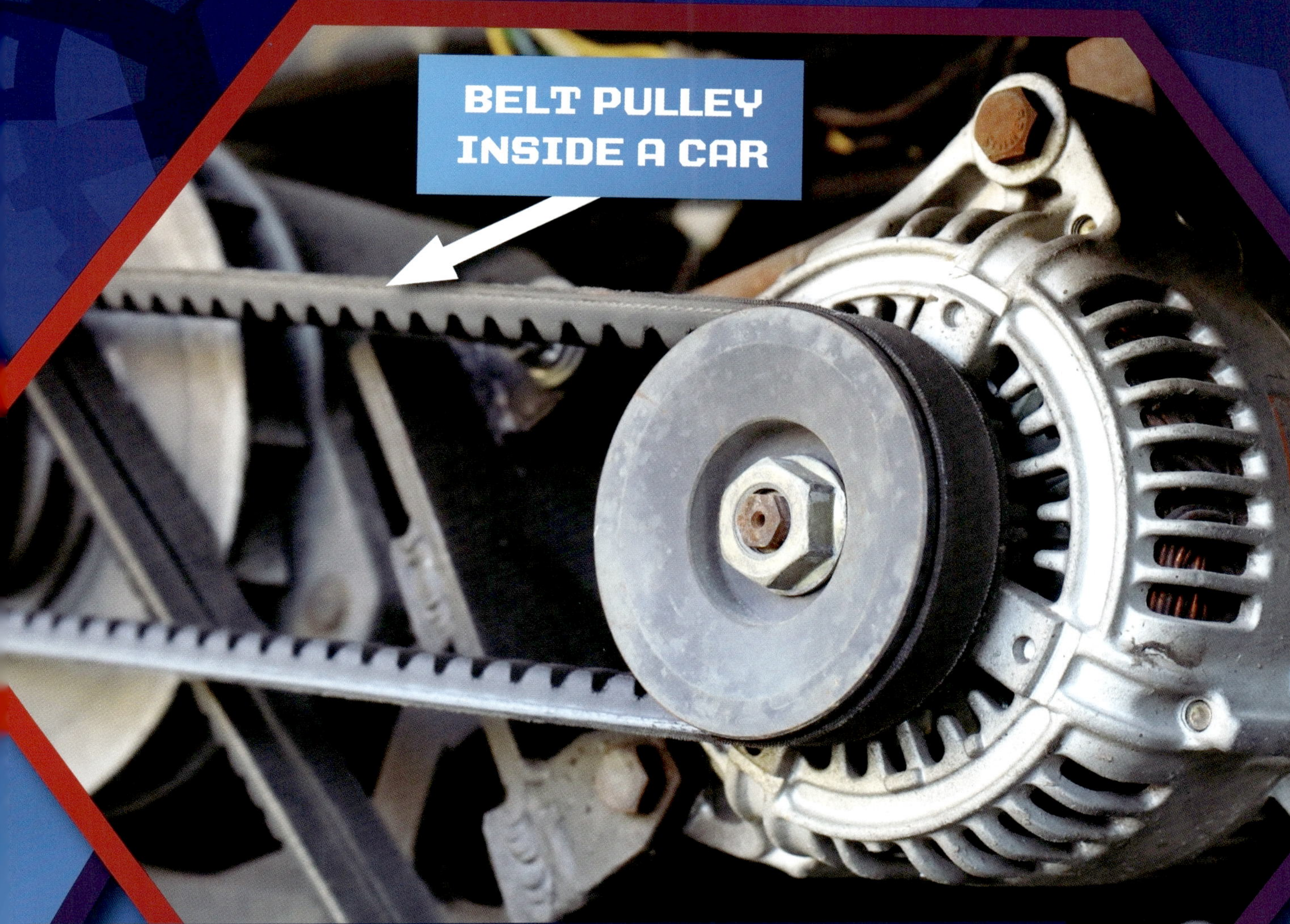

Pulleys found inside cars are called belt pulleys. Belt pulleys transfer movement from one rotating pulley to another. This helps power other moving parts inside the car. Most cars need belt pulleys to work.

Daily Pulleys

Elevators

Have you ever wondered how elevators get you to different floors in a building? Most elevators are attached to thick cables. They are lifted up or lowered down using a pulley system.

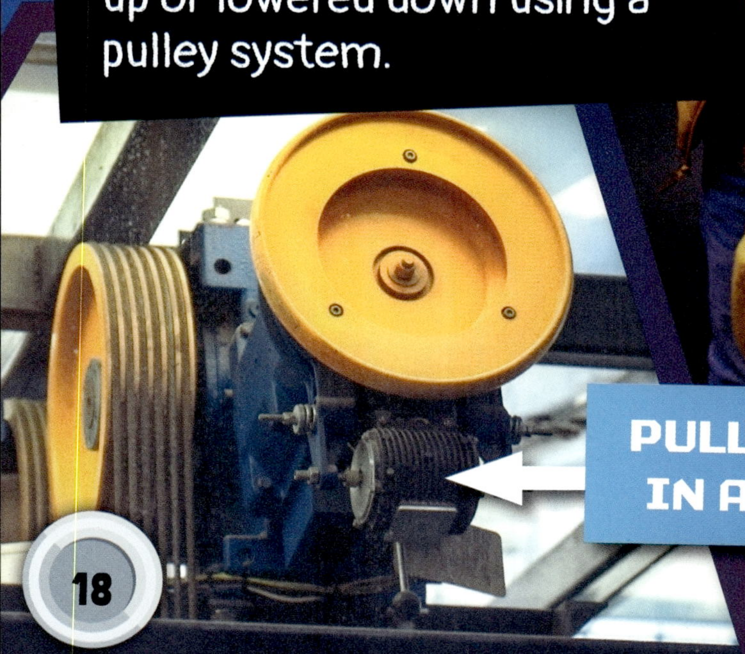

PULLEY SYSTEM IN A ELEVATOR

Blinds

Pulleys are not only found in big machines. They can be found in your house! Some window blinds have a small fixed pulley system. When you pull the string, the blinds open and close.

Ancient Pulleys

Wells

The well is a fixed pulley system that goes back hundreds of years. It is used to lift heavy buckets of water. Some well pulleys are lifted by hand. Others lift the buckets using other simple machines.

Drawbridges

Much bigger pulleys were used to lift and lower drawbridges. Drawbridges were built in castles hundreds of years ago. Lifting the drawbridge using a pulley system helped keep out unwanted visitors.

Super Simple

Pulleys are simple machines that are often used in big or complicated machines. However, they are also extremely helpful on their own and can be used for many different things.

We use pulleys every day, often without realizing it. Without pulleys, many other machines would stop working. Think about what you have done today. Have you used any pulleys?

Glossary

cylinder	a 3D shape with a top and bottom in the shape of a circle
force	a push or pull on an object
rod	a thin, straight bar, especially made from wood or metal
rotating	spinning around
structures	buildings or shapes that are made by connecting different parts
transfer	to move from one place to another

Index